H. Critchett Bartlett

The Digestion and Assimilation of Fat in the Human Body

H. Critchett Bartlett

The Digestion and Assimilation of Fat in the Human Body

ISBN/EAN: 9783337366315

Printed in Europe, USA, Canada, Australia, Japan

Cover: Foto ©berggeist007 / pixelio.de

More available books at **www.hansebooks.com**

THE

DIGESTION AND ASSIMILATION O

FAT IN THE HUMAN BODY.

AN EPITOME- OF LABORATORY NOTES ON PHYSIOLOGICAL AND CHEMICAL EXPERIMENTS BEARING ON THIS SUBJECT.

BY

H. CRITCHETT, BARTLETT, Ph.D., F.C.S.,

AUTHOR OF
ANALYTICAL PAPERS ON THE SUBJECTS OF FOOD AND THE NOURISHMENT OF THE BODY
IN " THE LANCET," " THE BRITISH MEDICAL JOURNAL,"
" THE MEDICAL PRESS AND CIRCULAR," " THE MEDICAL RECORD,"
" THE SANITARY RECORD," " PUBLIC HEALTH," ETC.

LONDON:
J. & A. CHURCHILL, NEW BURLINGTON STREET.
1877.

CONTENTS.

THE DIGESTION AND ASSIMILATION OF FAT IN THE HUMAN BODY.

INTRODUCTION.

Towards the autumn of 1872, a somewhat warm controversy sprung up between the late Dr. Edward Smith and myself, among others, respecting the proportional nutriment and digestibility of certain articles of preserved food, particularly in regard to "Australian meat" and "condensed milk."

The numerous letters which appeared in *The Times* and *Standard*, together with the more elaborate arguments brought forward in the columns of several scientific journals, attracted the attention of my esteemed friend and teacher, Baron von Liebig. A very interesting correspondence ensued, discussing minutely the various questions at issue.

Among other valuable results, I may incidentally mention the final repudiation by Liebig of the untenable assumption that his own "*extractum carnis*" was of a food value bearing any close relation to the nutriment contained in the whole bulk of meat from which it had been extracted. This candid admission of mistaken

views, which were previously advanced with no little
firmness and pertinacity, exhibited a great mind rising
superior to every self-interest and prejudice. As a direct
consequence of our intercommunication, this was naturally highly gratifying to me; but it is as an instructive
example, which may be borne in mind by all scientific
writers, no matter how distinguished their position, that
such a recantation should be regarded.*

During the progress of the discussion, Liebig expressed
a wish that I should place myself in communication with
Drs. Playfair and Bence Jones. The former was away
from London at the time, and when in town was neces-
sarily absorbed by the cares of high official duties; I therefore invoked the kindly assistance of the secretary of the
Royal Institution.

Dr. Bence Jones advised that a number of experiments
on the digestion of food should be undertaken; and, after
much consideration, wrote to me, suggesting the tabula-
tion of a very lengthy series of reactions, only to be attained by a course of investigation extending over several
years.

The various proximate principles of food were to be
administered without any mixture with other matters,
except water. The reactions to be recorded were as to
acidity, neutrality, or alkalinity during each stage of
digestion, from the mouth to the lower bowel. Not only
was the food mass to be thus tested, but my far-seeing
adviser was still more interested in obtaining similar
indications respecting the different conditions of the
various digesting juices. They were to be taken just as

---

* A precedent so frank was not lost on Dr. Edward Smith, who in his
later writings also virtually admitted that he had entertained erroneous
views even on the main points of the controversy.

secreted in their respective glands during the digestion of each single component of food, the like observations being registered before' and after digestional activity. Even beyond this, it was considered very desirable that the muscular tissues surrounding the digestive organs should be equally carefully tested, for reasons which I scarcely understood the important bearings of at the time.

While the processes of digestion in life were to be studied to afford the closest possible insight into the laws which govern the solution and absorption of the various food 'principles, the artificial digestion of single components of food, to be afterwards supplemented by simple combinations, was proposed to be experimented on in the laboratory with a completeness I have not yet been able to fully carry out.

Here was a programme ambitious enough, if affording any promise of leading up to a thorough comprehension of the true principles of the digestion of food, however complex in their alternations and combinations in the living human economy. Dr. Bence Jones was quite persuaded that such a course of experiments would contribute, at least, to the foundation of such knowledge, and he was eager to obtain the information to be acquired by this means.

I have so far quoted from such of his letters as I have still by me. Whether he would have considered his anticipations justified by the progress since made, I cannot presume to decide; but of this I am fully conscious, that in losing the benefit of his co-operation, advice, and encouragement, at his decease in the following year, many of the immediate scientific deductions logically to be drawn from the experiments made have been lost for ever.

Unable to devote the whole of my time and the considerable amount of money necessary to carry out the proposed research in its integrity, and being now hampered by the recent anti-vivisectional legislation, it is improbable that I shall be able to complete the physiological investigations commenced under such favourable auspices.

Accident, however, conduced to forward one particular train of experiments, namely, that on the digestion of fat in the living body; and as the artificial digestion of fatty matters was undertaken contemporaneously, the results have attained a greater advance than in any of the other sections. I therefore intend to discuss in this short treatise the general principles of digestion involved, to that extent only which may be necessary to explain their bearing in this instance.

For the last eighteen months I have had the valuable assistance of Dr. G. Overend Drewry, whose collaboration in the later physiological experiments has very materially helped to work out some of the more interesting problems connected with these peculiar digestional phenomena.

CHAPTER I.

A BRIEF ACCOUNT OF THE CIRCUMSTANCES, EXPERIMENTS, AND CON-
SIDERATIONS LEADING TO WHAT IS BELIEVED TO BE AN ELUCIDATION
OF THE DIGESTION AND ABSORPTION OF FATTY MATTERS.

The examination of the constituents of the gastric fluids
of the stomach forms naturally a leading feature of the
scheme proposed by Dr. Bence Jones. It was on that
score I was led to attempt the determination as to
whether the solvent power of pepsin upon the albumin-
ous portions of food is, or is not, accelerated and assisted
in its own function by admixture with other digestive
principles. This is broadly stated to be the case by
several American physicians; and the advantageous use
of the active principles of the sweetbread, or pancreas,
in helping the pepsin of the gastric juice in the stomach,
is vouched for by a chemist of high standing in New
York. My inquiries somewhat corroborate these views,
but in carrying out the trials I was for a time altogether
foiled in the endeavour to obtain in a convenient form
that portion of the pancreatic juice which should consti-
tute reliable pancreatin.

I could purchase useful pepsin, at a high price it is
true; and concordant results were obtained of its powers
of solution, even if the peptic solvent was almost
invariably below a fair standard. The pancreatin sold
at the shops, either as a powder or in a suspended state
in oil, proved, on the other hand, so unequal in its action
as to disturb all parity of experiment. If I had been
dependent upon any of the samples of prepared pancreatin
I was able to procure from our best known English

druggists, I should have been compelled to relinquish all further research, not only as to its supplementary action to pepsin, but in regard to its own specific functions.

Very little can be said in excuse for the manufacturers of the so-called pancreatin, who put forward, as the true active principle* of the pancreatic juice, preparations which rarely contain one-tenth part of the active principles to be found in the solid contents of that fluid. But what can be the word which will express the confidence in the want of knowledge of patients and want of testing by prescribers, or the unconsciousness of their own ignorance, which permits the sale of preparations termed pancreatin yielding less than 2 per cent. of any active principle of the fluid, and in some instances show none whatever? These last and utterly worthless samples have been more frequently sent to me than the unreliable preparations first alluded to, which do contain just sufficient of the active principles to cover the inertness of more than nine-tenths of their bulk.

At the outset, however, I determined to rely only on preparations made with my own hands; the hindrance occasioned by the bad pancreatin supplied from the shops was so far fortunate, inasmuch as it induced me to immediately devote a special attention to these experiments, instead of proceeding with them merely as supplemental to those commenced on pepsin.

Description of some of the Reactions and Functions of Pancreatic Fluid.

We could scarcely expect very close accord in the earlier observations respecting the reactions of the pan-

---

* It is noticeable that "the active principle" is always printed on the labels of the preparations in the singular.

creatic fluid. At a period of scientific research, however brilliant, when the dispute "raged furious" before the *savants* could decide the nature of the gastric acids, it is not surprising to find the pancreatic juice described as acid, acid-saline, neutral, faintly alkaline, and strongly alkaline. De la Boë and De Graf state it was acid to their tests and saline to their taste: Pechlin and Brunner compared the reaction to that of a neutral salt: Meyer discovered the fluid from the pancreatic gland of a cat to be faintly alkaline, which was corroborated by Magendie. Tiedemann and Gmelin, taking the first portion of fluid issuing from the pancreatic duct of a dog, which was opened for the experiment, found it to be slightly red and turbid; it was put aside without testing for some time. The next portion was whiter, with a bluish cast, and was decidedly alkaline, being considered the unmixed fluid from the gland. On reverting to the first portion, the test paper showed distinct acidity.

Baron Lucien Corvisart, the eminent medical adviser of the late Emperor of the French, referring to the previous experiments of Pappenheim, asserts that in one of its functions, namely, its digestion of fibrinous albuminoid matter, the action proceeds whether the fluid is in an acid, neutral, or alkaline condition.

With all the apparent contrariety of the reactions thus recorded, Corvisart's experiments are not sufficiently conclusive to hold good with regard to the reactions proper to the other functions of the pancreatic fluid. Leuret and Lassaigne, however, afforded the first reasonable explanation of the former seeming inconsistencies of reaction by proving that, when freshly exuding at the period of intestinal digestion, the fluid is always strongly alkaline in health: that shortly after its escape into any receptacle not protected from the air it becomes neutral, and after some time it turns acid.

Notwithstanding the valuable experiments of Bernard during the lengthy investigations he has bestowed on the subject, I propose to tread on almost virgin ground in the endeavour to identify with each peculiar function of the pancreatic fluid the characteristic reaction most suitable for the development of its activity.

In as few words as possible, I will give a short account of the different functions displayed by the combined principles contained in this complex digestive.

Bouchardat and Sandras demonstrated that raw starch, which remained untransformed in its passage through the gizzards of birds, and through the stomachs of those animals where the saliva is insufficient to change all the starch swallowed, is powerfully acted on in the intestine as soon as it is in contact with the pancreatic fluid. The corpuscles are eroded, dissolved, and transformed into sugar. Such digestion and chemical change involves the hydration of the starch by its taking up an exact equivalent of water; this is proved in the artificial action on starch by the pure active principle of the pancreatin when it is separated from its congeners, glucose being the result.

I find, when working with the pure principle of pancreatin which transforms starch, this action is favoured by distinct alkalinity at first, passing by degrees through the two other stages of neutrality and acidity. Sugar is formed out of the starch from the very first, and if the artificial transformation is conducted at a low temperature, not much above 65°, by the time the fluid becomes decidedly acid the chemical change will be effected without putrefactive decomposition. A heat equal to that of the body, when the pancreatic sugar-forming principle is not accompanied by antiseptics similar to those furnished by the gastric fluids and the bile, causes too rapid a change from alkalinity to strong acidity to convert the whole of the starch, before it develops large quantities of lactic acid, and putrefactive disorganization sets in.

My experiments included the use of many antiseptics to simulate in the laboratory the processes of nature in the living body. The result of employing the stronger antiseptics, in quantities sufficient to prevent putrefaction, was, I found, equally fatal to the active principles. Those of a lesser astringency, with certain exceptions, proved themselves less competent to delay putrescence than to arrest the functional power of the pancreatic principle. No foreign antiseptic enabled me to dispense with the natural preservatives of the bile and gastric fluids; but with these, or rather with a portion of these, as I shall hereafter explain, I obtained the transformations by the proper principles of the pancreatic juice at temperatures approaching that of the living body, and without putrefactive evidence.

The peculiar principle of pancreatin, exercising digestional power over nitrogenous substances, comports with the functions previously noticed by Corvisart, except that this principle, when separately used, instead of taking the concrete fluid for the purpose, is infinitely more energetic in a state of acidity.

Two special digestive actions of pancreatic juice having been separately watched in their somewhat opposite capacities, and the reactions most suitable to each individual principle ascertained, a third, by far the most essential of all the specific functions of the fluid secreted

by the pancreas, remains for investigation, namely, the
principle effecting the digestion of fat and oil.

As this portion of the pancreatic juice has not here-
tofore been even indicated, although its presence is
acknowledged at all hands, and as it forms the main
subject of this treatise, I must leave to the narration of
my experiments such description as may be the out-
come of long-continued labour to advance this subject.

Before entering upon detail, it is essential to be understood that, when I have used the word pancreatin, I have
so designated the combination of active principles contained in the pancreatic fluid, which may be presented in
a dry but perfectly soluble form. The word is in itself
objectionable, as it appears to refer to some single
principle, instead of being, as we have seen, composed of
at least three different active agents, each performing
distinct and even opposite functions. The term pancreatin
is, however, so widely accepted as representing pancreatic
matters soluble in water, insoluble in alcohol, and coagulable by heat, that it must still do duty for the crude
solids of the pancreatic fluid, which have been dried
without injury to their solubility or digestional action.

*Analytical Processes for the Separation of the Active
Principles and other Components.*

Reviewing the causes which in all probability contribute
to render the pancreatin of the shops so poor a digestive,
it occurred to me that the fluid taken in the first instance
may not have been procured when the glands were
secreting strong solvent principles. Several experimenters
have remarked that the maximum quantity and quality
of the pancreatic juice can only be acquired by removing
the pancreas of suitable animals immediately after death,

when digestion of food containing fat has commenced from three to seven hours previously. My own experience points to the superiority of the fluid obtained as it flows through the pancreatic duct at that period of digestion, but I found in practice a larger quantity is at once drawn from the gland, taken during vigorous digestion, if it is instantly subjected to maceration.

Difficulties were thrown in my way by the butchers and slaughtermen to whom I applied for sweetbreads from the animals under the necessary conditions. I was assured that all the pigs, calves, and other beasts killed for food are previously fasted so long that the upper intestines are always empty. This is regarded as of great importance by the butchers; but as, when fasting, the pancreatic glands only furnish fluids yielding the smallest proportion of active principles, the scruples of the slaughtermen had to be overcome.

At length I succeeded in procuring an occasional porcine pancreas, which I could depend upon being extracted from the carcase as I desired, but it could only be relied on in this respect when the chyle in the intestine was verified by ocular demonstration.

The pancreatic glands were passed through one of Nye's masticating machines until effectually pulped; this magma being exhausted by successive washings with distilled water at 40° F., was filtered at that temperature to prevent the least decomposition. The natural healthy secretion, as it flows through the pancreatic duct, contains about 9 per cent. of solids. These are composed of albuminoid matter soluble in water, albuminoid matter not soluble in alcohol, fatty matter, extractive matters and salts soluble in alcohol, together with other salts, chiefly of sodium. All these components were found in the pancreas, the fatty constituent being immensely larger than in the exuding fluid. The watery extract was then subjected to several methods of analysis to separate, as far as possible, not only the constituents named, but to attempt the isolation of the various active principles.

Organic matter dissolved in the watery extract is largely precipitable by alcohol, and to that extent, while retained in solution, is exceptionally liable to putrefy at temperatures above 40°. In this respect it resembles in a major degree some of the extractives of the saliva, but this unstable condition marks an essential divergence from the extractives of the gastric juice, which seem always combined in their first watery solution with an antiseptic which preserves them.

Such nitrogenous matter as is thrown down by strong alcohol resembles albumin as substance in composition. That which remains when the alcoholic solution after filtration is evaporated, and the residue washed with ether, is a pale yellow curdy substance.

Still more minute subdivision can be effected by other and more complex methods of precipitation; after which, the matters separated, though consisting of or containing the true active principles of the original fluid, can only be allied to ordinary albuminoid matter, as they do not react with most of the common tests for albumin.

Some of the more advanced foreign chemists have proposed several systems of very delicate separations, by which they claim to have produced two of the active principles of the pancreatic fluid, each being of great purity. The descriptions of the reactions said to accomplish these results having been submitted to me only by notes, certain discrepancies and a semblance of contradiction, so conveyed, might be removed, and the precise *modus operandi* rendered clearer if more minute details were afforded.

Acknowledging the advantage of having other minds working with the same object, I have not hesitated to modify the suggestions as they came to me, when in following them to the best of my ability I have not met with the expected results.

The aqueous solution of pure pancreatic fluid should be taken while strongly alkaline; or if the watery extract is filtered from the pancreas, the gland may be previously rubbed up with magnesium carbonate, to over-neutralize any acidity liable to be acquired during the long time necessary for the extract to pass through the filter paper. Into this, as my correspondent states, "the transparent colourless jelly, produced by mixing nitric acid sp. g. 1·5 with starch, is thrown. The instant it reaches the aqueous extract a white insoluble precipitate is formed, carrying down with it the active principle which digests fibrinous matter.* The solution can then be filtered off, and the insoluble precipitate redissolved in weak nitric acid, leaving the active principle pure." The possibility of success by this method evidently depends upon the correctness of the assertion that the principle carried down in the precipitate is unaffected by nitric acid.

Treating the aqueous extract in the same way, I substituted for the nitrated starch a solution of the less highly nitrated pyroxylin in ether-alcohol, which, while precipitating out with it the fibrin-digestive principle of pancreatin, leaves the starch-transforming agent in the

---

* The peculiar principle of the pancreatin fluid which digests nitrogenous matter is proportionally far more energetic in dissolving fibrin than albumin.

solution. This may be filtered off, and evaporated at a low temperature. The latter separation produces with certainty an almost colourless powder, absolutely inert as a solvent of fibrin, but a powerful starch converter. I greatly prefer the use of pyroxylin for the isolation of this particular principle, for I must confess I failed to reach this result by means of the nitrated starch.

If this method of analysis can be extended by redissolving the double precipitate, so that the fibrin-solvent is obtainable from a mixture of ether and water, it will greatly add to the value of the process, supposing we need not depend upon excessive nicety of manipulation. According to my own experience, however, I prefer extracting the fibrin-solvent free from the starch-transforming principle and all other extraneous matter by precipitating with calcic phosphate, which can be readily conducted with the necessary accuracy.

A naturally neutral watery extract should be taken from the gland ; that is to say, the usual alkaline extract is allowed to stand until it has become exactly neutral. Sufficient tri-basic phosphoric acid is thrown in to make up the whole to a solution of one in twenty. It is then slightly over-neutralized with a known quantity of lime in water ; to which the precise equivalent of phosphoric acid is added to produce an insoluble bi-calcic phosphate, from which the fibrin-solvent can be washed with distilled water.

Other methods of isolating these two active principles give results similar in all essentials to the solvents yielded by the foregoing analytical processes. The substantial truth of the propositions involved is thus confirmed, and we can now class these peculiar agents and their reactions as known principles, producing well-defined and characteristic organic change in those matters which are appropriate to each, and in no other kind of matter.

Recent research abroad appears to corroborate the earlier determinations of these experiments, but respecting the exact nature of the stimulating action possessed by such principles in effecting organic change, no general consensus of scientific conviction is yet attained. My endeavour is to deal with this portion of my subject as tentatively as possible; discriminating as far as may be between assumed facts, credible theories, and the uncertainties of both. My most important results must be regarded to some extent by the light which can be thus afforded; it is therefore essential to their explanation.

The Fermentative Nature of the Pancreatic Principles.

The stimulating action of the principles contained in the pancreatic fluid, as well as in all the other digestive secretions, is due to true fermentations produced by them. These ferments must not, however, be considered as strictly analogous to the alcoholic ferment of yeast; on the contrary, we must try to discern the extent to which their fermentative action differs from the better known, and therefore more widely acknowledged, ferments, so elaborately classified and described by Pasteur.

To illustrate more clearly the character of the ferments forming the active principles of the pancreatic and other digestive fluids, to those who have not made a special study of the wonderful phenomena attending upon the simplest and best known fermentation, namely, the alcoholic, I will explain the immediate connection between the fermentation by yeast and the fermentation by digestive principles.

Cane sugar is not fermentable as long as it retains

its composition is such—*vide* the text-books. Before it can be acted upon by yeast so as to be converted into alcohol and carbonic acid, a preliminary change must be first induced; this is effected when the cane sugar becomes hydrated and its composition is found to be made up of glucose (grape sugar) and lævulose (uncrystallizable sugar). The reagent producing this preliminary change was discovered by Berthelot in the water in which yeast has been washed, even after it has been filtered perfectly free from any yeast cells. From its peculiar action on cane sugar, it was termed inversive ferment; and Béchamp afterwards proved that it consists of soluble nitrogenous matter, either extracted from the yeast cells during their growth and reproduction, or excreted and thrown off in the process of their development.

The most remarkable characteristic of this soluble nitrogenous matter is the extraordinary rapidity with which a very minute quantity of it in solution causes the hydration of cane sugar. The peculiarity is strikingly indicated by all the soluble nitrogenous ferment principles excreted by the digestive organs; infinitely small quantities producing the most important and prolonged reactions.

The same fermentative hydration splits up cane sugar into glucose and uncrystallizable sugar, converts starch into glucose and dextrin; and the glycerides (fatty matters) become hydrated when split up into glycerin and fatty acids in the presence of water. The group of different albuminoid substances which composes the complex soluble nitrogenous ferments may also be regarded as the result of the breaking up of ordinary albumin by hydration. In this essential splitting up of

the various constituents of food, and in their consequent
hydration, we observe the true action of the inversive,
soluble, and digestive ferments.

Whether the inversive soluble ferment be taken from
yeast-washing, or any of the digestive ferment principles
are isolated from the other constituents of the digestive
juices, all soluble ferments are produced directly from the
living organism. As long as the yeast plant lives, the
inversive ferment is freely given off in solution, except
when the presence of special antiseptics arrests this
excretion without killing the plant. Similarly, the
digestive glands of the animal body may secrete fluids
in which the soluble ferments are rendered inert during
the period they are under the influence of counteracting
agents.

Boric acid at a certain strength of solution arrests the
vitality of inversive ferment and causes digestive ferments to remain dormant until they are washed free
from its control. Alcohol has the same effect, varying
only with the amount of its dilution. Citric, tartaric,
acetic and many other acids, usually found in a dilute
form in food, exert a very injurious effect upon the
digestive ferments, if they mix with the digestive fluids
while too concentrated or in excessive quantities.

Some volatile oils, on the other hand, exercise very little
subduing influence on the solvent powers of the digestive
ferments, although many of them, especially the turpenes
(represented by oil of cloves, lemon peel, etc.), are highly
antiseptic in the prevention of putrefactive decomposition. The volatile oils, which are themselves the
product of fermentation, appear rather to stimulate
digestion; of this we have a good example in the volatile
principle of mustard.

A wide line of demarcation is thus revealed between the direct ferments, complete in their own organism, like yeast, and the indirect soluble ferments which are the product of organism but are not so organized in themselves as to permit of reproduction ; and this is evidenced by the opposite reactions produced upon the two different classes of ferments by the antiseptics enumerated.

The Emulsifying Power of the Pancreatic Fluid.

Bernard was enabled to demonstrate that no other fluid secreted in the digestive organs, except that from the pancreas, can produce the complete digestion of a sufficiency of fats or oils. The remarkable property of forming an emulsion by mechanically holding the fat or oil in minute globules, with water filling up the interstices, is almost alone possessed by this juice. I am desirous of making an amendment to M. Bernard's statement that the power of emulsion is altogether peculiar to the pancreatic fluid, because I find that it is also produced to a very much smaller extent by the saliva. I am, however, completely in accord with the observation that the after saponification is only at present provable when oils or fats are subjected to the digestive fluids of the duodenum. Without refining too much on the possibility of a similar function perhaps appertaining to the saliva, it is evident that, as a very slight saponi-fying action is supposed to assist in conducting the fats of food through the walls of the absorbents, a still more minute agency of the same description may exist in the saliva without our being able to perceive it.

The state of emulsion is beautifully apparent if a drop from a sample of milk is placed upon the stage of the microscope. The globules are found floating about in that constant and rapid motion which takes place in all similar matters devoid of organization, and is denominated the Brownian movement. Without doubt, this movement in the globules is essential to the persistence of a mere mechanical emulsion; and although Mitscherlich and Moleschott were able to show that, in milk, this emulsive form is maintained by each globule of fat being coated with a thin pellicle of albumin, I have noticed the larger globules of butter, or those exceeding $\frac{1}{2400}$th of an inch in diameter, have a strong tendency to aggregate, on long keeping, by the rupture of their albuminoid vesicles.

The Brownian movement is not so manifest in the emulsion of the fats of food formed by the digestive fluids in the intestine, but the globules are even smaller than in milk. Emulsions are also sometimes formed which certainly do not contain any appreciable quantity of albumin, and these retain their form as long as others containing even a good deal of albuminose.

What is the peculiar principle, conveyed in the pancreatic fluid, which has so powerful an emulsive action? To this I can only reply that I have repeatedly found pancreatic fluid which both converted starch into glucose, and fibrin into peptone, but was extremely inert in the digestion of fat. In these cases, I have invariably noticed that the pancreatic glands were deficient in the very peculiar fatty constituents of which they are so largely composed when taken from a healthy animal.

The inference deducible from this fact is that the ferment principle producing instantaneous and permanent emulsion in the fats of food, is itself of a fatty nature, or is carried by or dissolved in fat or oil.

One other experiment corroborates this assumption, namely, if we take the other pure principles of the fluid absolutely free from fat, we find the wonderfully vigorous emulsifying property is almost destroyed.

A great misconception as to the real characteristics of a true pancreatic emulsion has been entertained by many, and but few appear to have studied the different aspects presented by such an emulsion as is produced on fat by the energetic action of pure soluble pancreatin, as contrasted with the coarse mechanical mixtures of oil or fat and water which are commonly supposed to represent this function of fermentative digestion.

Some seem to think that if a bottle of oil is shaken up with the compounds sold as the active principle of the pancreas, and a yellowish cloud is diffused for a time through the oil, an emulsion has been obtained. So it has, but not the true pancreatic emulsion, which forms an integral portion of the process by which fats are digested and assimilated. From the unvarying result of many hundred trials with the pure, active principles of healthy pancreatic fluid, taken at the time of digestion, I am perfectly convinced that no valuable result has been attained unless the emulsion formed is as highly refractive of light as milk. The colour may vary, according to the oil or fat used, from a far whiter fluid than the densest milk to the opacity and colour of Devonshire cream ; but unless at least the equivalent of the density of the best milk is produced in oil, when a third of water is held in suspension, no real pancreatic emulsion has been formed.

The mere mechanical mixture formed by common pancreatin is rarely better or more persistent than may be produced by rubbing up oil or fat with a solution of mucilage, or by a warm application of dissolved gelatin shaken with oil until it becomes cold.

The first essential towards the digestion of fat or oils in the human body is that it shall assume the state of the very finest and most permanent emulsion, and this is only known to be attained when the oil and water is perfectly opaque, from the minuteness of the globules. This is the first function of the pancreatic emulsifying principle, and by this alone can we be certain that it possesses its proper fermentative activity.

The Manner in which a New Principle is thought to have been Detected and Confirmed.

Proceeding systematically in the examination of the various separate ferment principles given out by the healthy pancreatic gland, when in the greatest activity, I arrived at the unavoidable conclusion that the reaction proper to the exercise of all their functions in combination is that of alkalinity. The alkaline fluid produces the perfection of emulsion described, but no chemical change is effected in the fatty matters of food by the pancreatic fluid alone, except after a lapse of time which it is inconsistent to suppose can be accorded in the intestine. Bernard believed that the slight saponification is perhaps commenced in the intestine, and then continued in the after processes of absorption. This latter proposition is not borne out by my experiments; the fat in contact with the pancreatic fluid, the bile, and intestinal juices in the bowel, being more evidently saponified than the oil in the villi, the lacteals, or in the blood. How far this may be assisted by the alkalies and other reagents of the bile and fluids from the intestinal glands, I will not wait here to discuss; merely recording my persuasion that the reactions induced by their means are of the very essence of the digestional changes of fat which enable it to be absorbed.

My attention being engaged more particularly in separating the principles and in identifying their peculiar functions, I had only arrived at the negative result before mentioned—namely, that the further chemical change of the fatty emulsion is probably not due to the pancreatic fluid alone—when, in manipulating the pancreas of a pig, I almost stumbled on what appeared to me a startling discovery.

After obtaining the watery extract from the gland, the strongest alcohol (not absolute) was added to precipitate the coagulable albumin. When this was poured off and evaporated, I found a peculiar greasy deposit. At first, I was disposed to attribute this to the fineness of the emulsion of fatty substance permitting it to pass through the filter paper; but on repeating the experiment, I obtained an almost bright filtrate of the aqueous solution, and the alcoholic extract became perfectly clear and brilliant. The deposit of grease was evidently not due to the fineness of emulsion, for the finer the particles of fat, the more opaque the milkiness would appear. I therefore treated the deposit with ether, carbon-disulphide, light spirit of petroleum, and other solvents of fats and oils.

Every successive test showed the ease with which the fatty portion was taken up by these menstrua, precluding the possibility of its being mainly composed of the glycerin liberated by the saponification of the fixed fatty acids forming the fat of the food previously given to the pig.

When I first saw the pig, it had been fasted for forty-eight hours; a meal of boiled potatoes mixed with beef suet was then given. Six hours afterwards the pig was killed, and the pancreas submitted to the processes alluded to. Yet here was a considerable quantity of a brownish oil, to some extent volatile, or at least diffusible at temperatures very little exceeding that of the body. Extremely soluble in a very small proportion of ether, it combined with the hydrates of sodium and potassium at the higher degrees of animal heat—a result worthy of note, as I am not aware of any previously known condition of change in the fixed fatty acids which produces soap at less than 176° F.

Control over the feeding of the pig for a period of fifty-four hours appears adequate to assure us that the glycerides of fatty acids, volatile at an animal heat, were not contained in the immediate constituents of the food recently given.

Absolute proof of this was desirable; so a portion of the same sample of suet was taken, treated with potassium hydrate and alcohol, and yielded a perfectly bright solution of soap. It was then decomposed with a mixture of hydrochloric and lactic acids, these being found in the stomach of the pig. After well washing with boiling water and drying in vacuo, 95·6 per cent. of fixed fatty acids remained, leaving, if the due proportion of glycerin be reckoned, no room for any trace of glycerides of volatile fatty acids in the suet before the pig ate it.

The obvious conclusion to be drawn from this extraordinary result is that the animal digestion of fats and oils renders a certain portion of the fixed fatty acids soluble in water and alcohol, and in a mixture of both. If this is correct, as is attested by all my after experiments, the process of absorption of fat through the membranes of the villi of the intestines and the other absorbents is capable of another explanation besides that doubtfully assigned to the slight saponification also observed.

As a short summary of the results of these preliminary experiments, I may state :—

Firstly, that the formation of even the most perfect emulsion of fat or oil with water does not of itself render any portion of the oil or fat soluble in water, nor does the fat or oil take up any water, except mechanically, to become, so to term it, hydrated.

Secondly, the addition of alkalies to fixed oils or fats at common temperatures, whether the alkalies are as hydrates or salts, does not affect such transformation.

Thirdly, mixtures of fixed oils or fats with water and the solids of the pancreatic fluid, when kept for long periods, promote the separation of the fatty acids from the glycerin. This decomposition is, however, effected by that most disgusting butyric fermentation, which is set up in the presence of putrefying nitrogeneous matters supplied by crude pancreatin, and is never found in a healthy state.

Fourthly, saponification is not the only natural solution of fatty matters in water at the temperature of the body.

Fifthly, that fatty matters are found in the pancreatic fluid and in the pancreas in a free state; in solution with water (hydrated); and are also slightly saponified in the intestine. These, or some of these different forms of fat, are distinguishable in combinations of various proportions in the intestine, the pancreatic gland, the thoracic duct, and while the fat or oil is passing through the walls of the absorbents.

CHAPTER II.

A SLIGHT SKETCH OF THE PHYSIOLOGY AND CHEMISTRY RELATING TO THE PRELIMINARY DIGESTION OF FATS.

WHEN fat or oil is taken in food or with food, as I am led by the whole course of these investigations to believe it always should be, the first introductory step towards its digestion is effected in the mouth by all animals possessing the instinct to thoroughly masticate their food. I am aware that this is contrary to the generally received teaching, which only acknowledges the chemically digestive action of the saliva on starch. In a subsidiary degree, the valuable results of good mastication are admitted, so far as appertains to the preparation of the nitrogenous portion of food for the stomach. I wish to claim the possibility of something more than this being effected during tne chewing, mixing, and incorporating to which the separate fatty matters taken in an ordinary meal are subjected prior to being swallowed. The presence of the alkaline ferments of the saliva produces a coarse emulsion, which, when the other portions of food are easy of chymification, is not altogether destroyed by the action of healthy gastric juice. Observations founded upon experiments with dogs do not enable this to be properly shown, as these animals seldom masticate food of this character.

Many years ago, Dr. Wright announced that the degree of alkalinity of the saliva during digestion was in direct proportion to the acidity of the stomach fluids. Of the precise accuracy of this I am by no means convinced, although I must endorse Dr. Bence Jones's observation

that, during the excretion of acid in the stomachic digestion, the alkalinity of the other fluids is increased. I
take it that this means the total alkalinity of *all* the
other fluids is increased to the extent of the acid with-
drawn from them by the local digestion in the stomach.

The variation of acidity in the gastric fluid certainly
produces a marked difference in the appearance in the
stomach. In the case of a meal containing hard, indiges-
tible, or irritating matters, the peptic solvents are found
highly charged with acid, and in the absence of sufficient
saliva, the fatty substances are seen separated from the
chyme, but pass through the pylorus at the same time in
a melted or oily form.* If the other components of the
meal are easy of chymification, the emulsion of fat by a
full quantity of healthy saliva is generally sufficiently
maintained against the feebler acids which are furnished
by the stomach to undertake the lighter task of trans-
forming the freely digestible albuminoids into peptones.

Fats or oils taken into the stomach without any other
food are usually swallowed rapidly ; they are, therefore,
less acted on by the saliva. The after flow of saliva
is also less abundant; this may result from the absence
of the muscular play upon the glands incident to the
masticatory action, or from the want of that sympathy
which excites the outpour of this fluid after the savour
of the other principles of food has been appreciated by
the palate. Be this as it may, fatty matters or oils, taken
by themselves into the empty stomach, are less emulsified
by the ptyalin of the saliva and the oral mucus, and
are more distinctly separated by the stomach juices than

* When the acidity of the stomach is provoked to excess, no amount
of alkalinity which can be conveyed by the saliva will obviate the
separation of the fats.

when they form part of mixed meals. In the same way, the peculiar emulsionary function of pancreatin is rendered nugatory when oil or fat is not accompanied with other solid food. A fully pancreatized oil or fat may sometimes pass through the stomach by itself without being thrown out of emulsion, and in this form, if there is a normal flow of bile, it may be, and probably is, rendered easier of digestion. The reactions of the stomach juices are, however, almost invariably acid enough to separate the oil from the emulsion if no other food is present to absorb the gastric fluids.

By tying the pylorus before any portion of the mixed contents of the stomach can pass through, the progress of chymification and stomachic digestion may be watched with advantage. The pultaceous mass is not found to be uniformly acid, neutral, or alkaline. For the first hour or more, different portions exhibit these several conditions. Neither the flow of alkaline saliva nor of acid gastric juice is altogether continuous, observation leading to the conclusion that their influence on some of the food in the stomach is frequently alternative. The maintenance of fatty emulsion, however, depends almost entirely upon the alkalinity retained in it being at least equivalent to any acidity with which it comes in contact.

After fasting for forty-eight hours, fat or oil may be kept in the stomach for several hours by closing the pyloric orifice, without any increase being found in the trace of fat usually contained by the surrounding lymphatics. The addition of a very small quantity of bile and pancreatic juice to the fat in the stomach, however, enriches the lymph taken from the thoracic duct in the absence of chyle, until it yields nearly one per cent. of fatty and other matters soluble in ether.

This would appear to show that a very limited but rapid digestion of fat may, under certain circumstances, take place in the stomach, and that it may be absorbed there and enter directly into the circulation. The probable capacity of the lymphatics for carrying fat may perhaps be claimed from the comparative analysis I was enabled to make of the vascular glands of two hedgehogs; one killed just before hibernation, the other being reserved until after that period. Both were of the same age, and as nearly as possible of a size. The thymus and appended fat glands, connected only with the blood vessels and lymphatics, and with no direct communication with the lacteals, became literally overcharged with fat before the winter sleep, and 67 per cent. of fat in the dried substance was extracted with ether. Whereas, after hibernation, only a bare trace of fat was discoverable in the similar glands of the sleeper when first awakened by the approach of spring.

The distribution of fat by the lymphatic system, if this can be admitted, affords a slight clue towards assisting our comprehension of certain otherwise unaccountable physiological facts.

It is well known that the body of a well-nourished adult contains nearly one-fourth of its weight of fat. Equally well ascertained is the chemical truth that the entire blood in circulation does not yield one-thousandth part of the fatty components of such a body.* The natural inference would therefore be that the elimination, use, and waste in the human economy of life must be particularly slow and small in quantity. The exact reverse is, however, the case, as in starvation or wasting

* A man of eleven stone should have about 28 lbs. of fat and 12 lbs. of blood in his body, which contains about ½ oz. of fat.

of the bodily tissues, the whole of the fatty structures are reduced more rapidly and to a greater extent than in any of the other tissues or juices.

Before death from want of food occurs, as much as 90 per cent. of the bodily fat is starved out; and it is a noteworthy corollary to observe that, next to the fatty tissues, the pancreas, salivatory, vascular, and other glands lose the largest proportion of their substance, namely, 86 per cent., from the same cause; these include the whole of the fat-storing and digesting organs.

If it is difficult to understand how the blood can convey all the fatty matters to their destinations so as to supply the large and rapid daily waste, is it not feasible to believe that the direct circulation of the blood may be supplemented by the fat-carrying capacity of the lymphatics?

However small the power of digesting fat in the stomach may be, it seems likely to afford, artificially, the most rapid means of recuperation after great wasting of the fatty tissues; assisting the resuscitation of functional activity of the natural fat-digesting organs. Certainly the main hope of nourishing the pancreatic and other digestional glands, so that their specific powers may be stimulated, depends upon the conveyance to these organs of that constituent fat which they are unable to take up for themselves.

To assist this supplementary action of the absorbents, or, rather, to enable lymph to take up fat at all, the one absolute necessity, apparent to me by direct experiment, is the perfect solubility of a portion of the fat or oil in the watery constituent of the lymph, minute though it be.

Pure lymph, as we know, is quite clear and devoid of milkiness, even to the minimum extent indicated by opalescence. After fasting from food for some time, the thoracic duct contains transparent lymph only. Yet, when the pylorus is tied, previous to a meal being given in which fat is contained, the increase of fat to be found in the thoracic duct does not cause, in the slightest degree, the characteristic milkiness of chyle absorbed through the lacteals. In other words, this lymphatic absorption of fat, which is proved by analysis, consists only of fat soluble in the watery lymph.* This leads me forward to the consideration of the manner in which any portion of fat becomes truly soluble in water so that it may be absorbed.

CHAPTER III.

THE DIGESTION OF FAT IN THE SMALL INTESTINE.

WE are obliged to acknowledge that, with the highest microscopic powers, we fail to find the faintest indication of any pores or ducts in the recognized absorbents of the intestine. The absorption of food by blood vessels is apparently almost omnivorous, many substances in a sufficiently subdivided form seeming to be taken up by them ; in addition to which, solutions and gases of most kinds are also freely received and mingle with the blood, frequently without being assimilated.

* I am particularly anxious to avoid laying undue stress upon the amount of fat digestible in the stomach or capable of being carried by the lymphatics. The first is, I am aware, but very small, the latter being difficult of estimation.

The penetration of solid particles, such as the sharp dust of charcoal, cannot, however, be regarded as absorption, being due to attrition forcing them mechanically through the walls of the vessels. The true absorption of liquids also depends upon their suitability to mix with and afterwards form a portion of the blood; and I have proved by repeated experiments that no other kind of fluid enters naturally into the circulation, although it may be forced into it.

Fat is found in an average sample of blood to the extent only of 1·5 per thousand, and this partly in a soluble form and partly as serolin. Free, fixed, or uncombined fat will neither mix with to form part of the blood, nor will it in this condition pass inwards through the walls of the vessels. A certain amount of a different kind of fat is, however, retained in the blood, which only becomes insoluble after exposure to the air; this is probably excrementitious, or a residue accumulating from the soluble fat which is precipitable from solution by the salts of the alkaline earths, and particularly by phosphate of lime. It resembles cholesterin, but melts at a much lower temperature, namely, 97° F. This peculiar fat is, I find, eliminated through the sebaceous glands, as well as by the bowels.

The chief supply of fat to the blood is not by direct absorption from the blood, but from the chyle taken up by the lacteals. During the active absorption of chyle containing fat, the villi become whiter and more opaque, and when the fat has been duly prepared by admixture with bile, pancreatic fluid, and the juices of the intestinal glands, a portion of this fatty constituent may be traced in the investing epithelium. The columnar cells, here and there, become filled with brilliant globules of oil, which I have ascertained to be free oil, containing no water or other constituents of the chyle.

The columnar cells are so small that 1600 of them placed end to end only measure one inch, their diameter being less than half this size; the value of the most perfect emulsion is therefore evident, by which only minute globules are presented to the cells. But a mere mechanical mixture of oil and water, however finely subdivided for the moment, does not enable the oil to permeate the delicate membranous walls of the cells. The experiments proving this have an importance which demands a more detailed description than my present space permits, but I hope shortly to further particularize.

After fasting a dog for two days, a meal of beef kidney fat was given, in which were a few pieces of old tanned leather to excite due peristalsis. The bile and pancreatic ducts were previously tied, and a ligature was passed round the bowel, about two feet from the pylorus. At the end of three hours the animal was killed, and the condition of the intestine was examined, both microscopically and chemically, for any fat it might have absorbed. Scarcely any oil globules were found to have entered the cells of the villi with which the fat had been so long in contact, and by exhaustion in boiling ether no difference could be perceived in the amount of fat extracted from the portion above the ligature, as compared with that obtained from an equal weight of intestine taken from below the part closed by tying.

A similar experiment was performed upon the intestine of another dog, the only difference being that an alkaline solution of pancreatin with bile extractive was mixed with the fat before it was swallowed. A third dog was fed as in the first case, but the natural secretions of bile and pancreatic fluid were permitted to flow into the intestine. In both these latter cases the fat, in an oily state, was found in the cells of the villi, and analysis gave evidence of a very large absorption of fat in the parts above the ligatures, while in those portions beneath it no increase of fatty constituents was yielded.

These results are in themselves sufficiently conclusive, but I have observed that they are corroborated by all I have ascertained concerning the still more interesting processes by which fats pass through the various membranes and tissues, when suitably prepared either by natural or artificial means. It is at the very point of transmittance that the complex actions and reactions occur which are included in the but little understood digestional absorption of fat.

The moistened membranes of the villi, lacteals, and blood vessels do not pass free fixed oil by endosmosis, as is believed by some ; neither will mere alkalinity assist in its absorption, as I was able to demonstrate by a fourth experiment. The mixture of pure pancreatic fluid and fat appears almost equally incapable of being taken up by the absorbents, but the experiments to determine this were not quite so conclusive, probably because we were unable to prevent the presence of fluids from the intestinal glands. Many and various supplementary examinations confirm these data, and the necessity is plainly shown for the presence of an alkali, pancreatic fluid possessing full fermentative vitality, and certain elements of the bile, to render fat truly soluble in the fluid before it can be absorbed.

A very minute portion of the soluble oil appears sufficient to effect the transfusion of a large quantity of a fine emulsion. It may act by rapid endosmosis, carrying with it the fixed fats, and may then return by exosmosis, having deposited the globules of free oil within the membranes of the cells. By a similar action the fixed fats may pass from one membrane to another, until it is mixed with the chyle in the lacteals and the blood in the capillaries of the bowel. I am not in a position to demonstrate that this *is* the precise manner in which the soluble portion of the oil enables the other portion to be absorbed, but I have proved that fat which contains no soluble glycerides is not absorbed until the reactions of the bile and pancreatic fluid have rendered a portion of it soluble in water.

Whatever description of fat may have been eaten, it must be so far transformed as to approach in composition to that of butter or the fat of milk which has passed through the mammary glands.

It may not be generally known that butter is, to a certain extent, easily rendered soluble in water; but as this peculiarity affords the distinctive difference between pure butter and the common fats with which it may be adulterated, and as this is now relied on in butter analysis for the detection of such adulteration, I may, perhaps, be excused if I repeat the evidence of this instructive fact, which first dawned upon me some three years back.

Ordinary mutton, beef, and pork fats are composed almost exclusively of the glycerides of the fixed fatty acids, such as the stearic, palmitic, and oleic acids. If these are saponified with hydrates of the alkalies, and the soap is decomposed with a dilute acid, such fats will yield more than 95 per cent. of the fixed fatty acids, which will float upon the water, being absolutely insoluble in that condition. If, however, butter is saponified, nearly 14 per cent. of other fatty acids and glycerin are set free; these are both volatile and truly soluble in pure water.

The analogy is perfect between butter and the fatty matters of food, after they have been acted on by the bile, pancreatic, and other essential fluids in the intestine; a portion of the fat, varying from 4 to 7 per cent., being in true solution by the time its digestion is complete and it is ready for absorption. A slight saponification is evidently required to form a hydrate of the fatty matters by the fixation of a portion of water. This product is in its turn decomposed, and the soluble fatty acids and glycerine liberated to enter into solution. Bernard has always maintained that soap is formed in the intestine, and I am sure that no careful experimenter can fail to find it during a vigorous digestion of fat; but I diverge from M. Bernard's views when he assumes soap, as such, to be the vehicle—much more, as the only vehicle—for the transfusion of fats through the various membranes.

As I before stated, I am not greatly concerned now to adopt any dogmatic theory of the exact *minutiæ* of such transfusion, but I must emphatically declare that I can find no actual soap except in the intestine, even when a considerable quantity of soap has been injected into it. I therefore lean strongly to the opinion that saponification is only a preliminary process, confined to the bowel alone. The soap, when formed, has to be split up before its fats are absorbed through the several membranes so that they may be taken up in the circulation.

The very support afforded by this possible advance upon previous tenets enhances the practical value of the discovery that soluble fatty matters are essential to the healthy secretion of the pancreatic glands. The experiments which prove that the active functions of the pan-

creas depend chiefly, if not entirely, on its being itself
supplied with fat soluble in water, seem incidentally to
point out the conditions under which the supply can be
afforded.   Having taken the subject thus far in recording
a brief generalization of my rough analytical notes, I
must reserve a few words explanatory of the attempts
made to arrive at a synthetical product complying to
some extent with these essential conditions.

CHAPTER IV.

ARTIFICIAL AIDS TO THE DIGESTION OF FATS.

ALL animals suffering from emaciation labour under the
same degeneration of the fat-digesting organs in varying
degrees.   From whatever original causes the pancreatic
functions have lapsed into abeyance, no sufficient or
healthy flow of the digestional fluids is ever found during
wasting diseases.   This is provably true of dogs, pigs,
calves, and other animals which become attenuated when
plentiful food is provided.   It is not difficult to insert a
drainage tube into the pancreatic duct and test the difference of the emulsifying power of the fluids obtained on oil,
which may be compared with that produced by animals
possessing an obviously healthy digestion.   A regular and
reliable flow of bile is equally essential to the formation
of soluble fat or oil, and this is as frequently found want-
ing during the digestion of food by all animals losing
weight, unless it be from over-exercise or want of proper
food.

Nothing appears to restore the healthy functions of the liver and pancreas in these cases, except the frequent ingestion of oil or liquid fat, so treated artificially that it is already partially transformed by fermentation and the reaction of bile. Seized on with avidity by the absorbents, it is insensibly assimilated by the digestive organs, until they gradually become strengthened, not only to provide their own nourishment, but to transform a sufficient quantity of fat to supply the inevitable waste throughout the body.

The fat or oil most suitable for general nourishment is evidently that which most nearly approaches the composition of the fat to be renewed, but a fallacy underlies the proposal that the small quantity necessary to give a periodic impetus to the digestion of the common fats of food need be, or indeed ought to be, of this exact composition.

It is admitted by those who contend for the administration of the more solid fats as a pancreatic emulsion that, in the first instance, oil, such as cod-liver oil, " can be hurried most rapidly into the *pulmonary* (?) circulation ; it is the fluid oleinous kind of fat that can pass by the portal instead of by the lacteal route." It is, as Dr. Dobell says in another place, " like water to the uprooted flower." But, then, this candid writer proceeds to advise the use of solid pancreatized fat, because " if you keep it (the flower) in water after it has revived, instead of planting it in good soil, it will droop again and die for want of materials on which to live." There would be great weight in this, if the fat proposed to be made into the emulsion is exactly of the composition of human fat, and no other fat should be taken in the ordinary diet. Dr. Dobell, however, advocates the use of a far more solid

fat than human fat, and forgets that the animal fats of
food contained in his own dietary are also of this precise
nature. He also loses sight of the obvious necessity for
a due admixture of the "fluid oleinous kind of fat," to
approximate the harder fats of the diet to the normal
fatty matters of the human body. To continue his own
metaphor, the flower requires not only "good soil," but
periodic watering.

The advantage of an emulsion of the more fluid oil to
temper down the too great solidity of the other fats taken
in ordinary diet is therefore manifest. As I have proved
this to be the case with pigs, in which the lard more
nearly approaches the consistence of human fat, I think
we may assume the same to hold good in the human
digestion of fat.

Taking perfectly soluble pancreatin,* and completely
emulsifying a suitable oil with water (two parts to one),
I find there is a difficulty in preserving the ferment
principle from working itself out in the course of a few
days; after which the pancreatized oil will not com-
municate its emulsifying property to other fats or oils
with which it may be brought in contact, as it does when
the ferment is still in vigorous activity. Hence, we
should require a fresh preparation to be made almost
every day in summer-time, or the fatty matters of food
will not be transformed so as to be digested except by
those who do not require such assistance. To obviate
this, I made numberless experiments with temporary
antiseptics, and I conclude that one only is really suitable.
Boric acid appears to arrest any after fermentation in the

* If any portion of the pancreatin is insoluble in water, it denotes a
highly objectionable mode of preparation ; the true ferment being killed,
and the whole exceptionally liable to ammoniacal decomposition.

emulsion without injury to it; and when it is combined with the soda to represent that constituent of the bile always forthcoming in the naturally healthy digestion of fat or oil, I observe that the salt formed becomes so dissolved and diluted in the digestion of food that the pancreatic ferment resumes its activity, and all the other fats of the meal become in a like manner transformed.

At this period of liberating the ferment from the temporary antiseptic influence of the boric acid, there is a liability to a slight putrefactive decomposition, which is only restrained naturally by other principles of the bile. At first I was led to attempt this artificially, by adding glyco-cholic acid in its original combined state (the "crystallized bile" of Plattner), but the flavor was so nauseous that I could not get animals to swallow oil prepared with it. Reflecting that the pancreatin used was from the pig, and, according to Strecker, the bile of that animal contains glyco-cholates differing from ox bile, I refined the glyco-hyocholic acid until the objectionable bitterness was removed, and I was pleased to observe that its function in the intestines was but little impaired.

Testing oil prepared with soluble pancreatin, soda, boric acid, together with a trace of hyocholic acid, I have every reason to believe that a transformable modification of the oil is reached, which is digestible in the most atrophied condition of the organs.

All the elements for a gentle but rapid saponification are insured, and the splitting up of the soap is favoured by the presence of a small quantity of already hydrated oil in solution. How little of this actually soluble glyceride of volatile fatty acids is sufficient to continue the hydration of the remaining fats during digestion is almost impossible of estimation. I have, however, satisfied myself that very small proportions yield good results, but slowly; a larger quantity promoting extreme rapidity of absorption.

Nearly two years of almost incessant observation affords a fair means of judging as to the corroboration since given by constant repetition of the more important determinations. But the familiarity with the indications thus acquired may inspire more confidence in the results than might be accorded to the analogy between experiments upon the lower animals and the effects of the same kind produced in the human body. As soon, however, as the principle involved was made plain, every opportunity was embraced of watching the direct working out of the problem, as applied to the assistance it affords to the digestion of fat in the wasting diseases which afflict humanity.

This more properly belongs to the science of medicine than to that of pure physiological chemistry. I was, therefore, fortunate in obtaining the assistance of Dr. Drewry in elaborating the use of fatty matters containing hydrated and soluble oil in cases of consumption and other forms of wasting in the tissues. His testimony as to its adaptability must speak for itself; but some of his cases have been within my own frequent observance, and I cannot refrain from expressing the intense gratification it has afforded me to note the almost immediate gain in weight and improvement of health which has resulted, even in some of the worst instances.

ADDENDUM.

EARLY EVIDENCE OF THE COMMENCEMENT OF WASTING DISEASES DISCOVERED BY THE EXCRETION OF SOLUBLE FAT.

URINE and fœcal excreta have been frequently submitted to me for examination by members of the medical profession, in cases where there was reason to believe a direct loss of fatty substance occurred through the kidneys or bowels.

In certain morbid conditions I have found free fat or oil in the urine. It was not detected in most cases until after cooling, when it assumed a chylous or milky appearance. This usually happened without the presence of any considerable quantity of albumin. In other cases fat or oil was found which never showed itself to the eye, and these were invariably connected with phthisis, tabes, or nervous wasting. The quantitative estimation of these transparent fats always presented difficulties I could not account for. In drying the residues I experienced a loss of weight which continued as long as they remained in the drying chamber of the water bath; repeated weighing could not, therefore, be depended on to confirm the absence of the solvents. Since making the discovery of the soluble and volatile fatty matters produced by fermentation in the intestine, this discrepancy is explained; and I have no doubt, from samples more recently analyzed, that mere traces of soluble fat in the urine may frequently mark the earlier stages of many wasting diseases. To obtain a certain verification of this slight excretion, vapor distillation must be resorted to, and the most accurate manipulation is required to prevent the loss of such traces of soluble volatile fatty matters as are sometimes to be found at the commencement of the disease.

Similarly, I now always submit the fœcal matters to this, among other delicate tests. The result is that a false or secondary digestion of fat is often found to have taken place in the lower bowel without any benefit being derived from it. On the contrary, it seems to denote one of the first symptoms of the degeneration of the natural fat-digesting organs. As this is of importance in pointing out a possibly unsuspected mischief, I have thought attention should be directed to the means analysis affords of confirming or removing uncertain suspicions as to such morbid conditions. These may, or may not, be intimated by a slight glistening film, either on the surface of the solid excreta or floating in the urine.

Interesting as such investigations are in supplementing the foregoing inquiries, I have had to regret the interruption lately of experiments extending over nearly seven years. The unexpected enforcement of certain rules of Gray's Inn has practically closed my laboratory there for these purposes. I have, however, now made special arrangements at my new laboratories in Duke Street, Grosvenor Square, which will, I hope, enable me to complete at least some of the other physiological experiments, and to proceed with analyses such as have been lately forbidden to me.

25, *Queen Anne Street,*
    Cavendish Square.

www.ingramcontent.com/pod-product-compliance
Lightning Source LLC
Chambersburg PA
CBHW022029190326
41519CB00010B/1635